6th Grade Math

Volume 5

© 2013 OnBoard Academics, Inc
Newburyport, MA 01950
800-596-3175
www.onboardacademics.com

ISBN: 978-1-939796-98-1

Table of Contents

Solving Equations

How many cows are in each group?

Farmer Pam has **32 cows**. She splits them up into two groups. One group will remain at the farm for milking, while the other group will be sent to market. The group for market has **6 more cows** than the group for milking.

How many cows are there in each group?

List the key information from the paragraph above.

1.

2.

3.

Write an equation

Key information:
- 32 cows
- 2 groups (milk and market cows)
- 6 more market cows

Relationship: [Milk cows] + [Market cows] = 32

Let **c** represent the number of milk cows:

Equation: [c] + [   ] = 32

Solving using algebra.

Study the illustration below to discover how to solve the previous problem with algebra.

	Write an equation	$c + (c + 6) = 32$
	Simplify	$2c + 6 = 32$
Why?	**Subtract 6 from each side**	$2c + 6 - 6 = 32 - 6$
	Simplify	$2c = 26$
Why?	**Divide each side by 2**	$\dfrac{2c}{2} = \dfrac{26}{2}$
	Solution	$c = 13$
	There are 13 milk cows and 19 market cows.	

Write an equation.

Farmer Pam has a total of 20 chickens in 3 chicken coops. She has an equal number of chickens in the first 2 coops, but there are 4 fewer chickens in the remaining coop

How many chickens are there in each coop?

Complete the equation below. Let 'c' represent the number of chickens in the first coop.

$$c \ + \ \boxed{} \ + \ \boxed{} \ = \ \boxed{}$$

Stretch your knowledge.

For helping out on the farm over the vacation, Farmer Pam gives each of her children a dollar amount equivalent to their age in years. This vacation, she gives 38 dollars in total.

Henry is six years older than Arvin, and Arvin is two years younger than Ernie. Write an equation for the amount of money each child receives in terms of x.

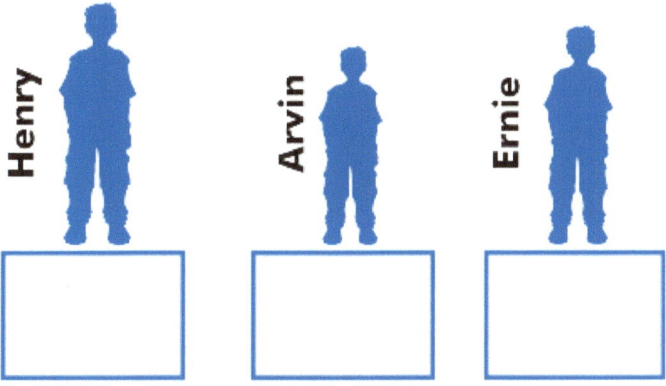

Let x represent Arvin's age.

How much money does each child receive.

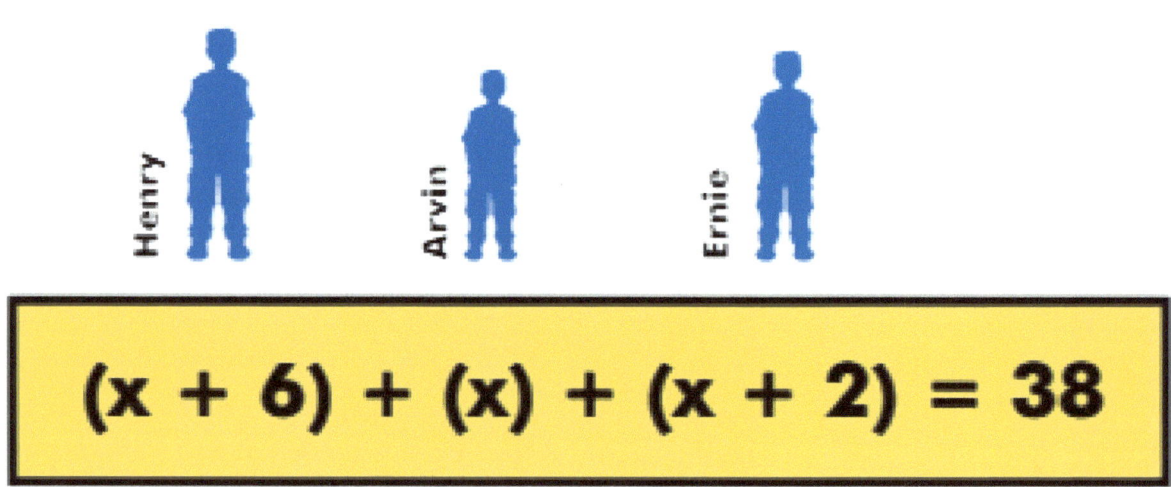

$$(x + 6) + (x) + (x + 2) = 38$$

Name_____

Solving Equations Quiz

1 An expression is the same as an equation.

2 Ernie is twice as old as Grace. Grace is x years old. The sum of their ages is 15 years. Which is the correct equation?

    A $3x = 15$

    B $x^3 = 15$

    C $x^2 = 15$

    D $x(x + x) = 15$

3 $3x + 5 = 23$ $x = ?$

4 $x + x - 9 = 31$ $x = ?$

The Coordinate Plane

Key Vocabulary

coordinate plane

ordered pair

quadrant

x-axis

Study the graph and its labels below.
Note the x and y axis, the quadrants and the origin.

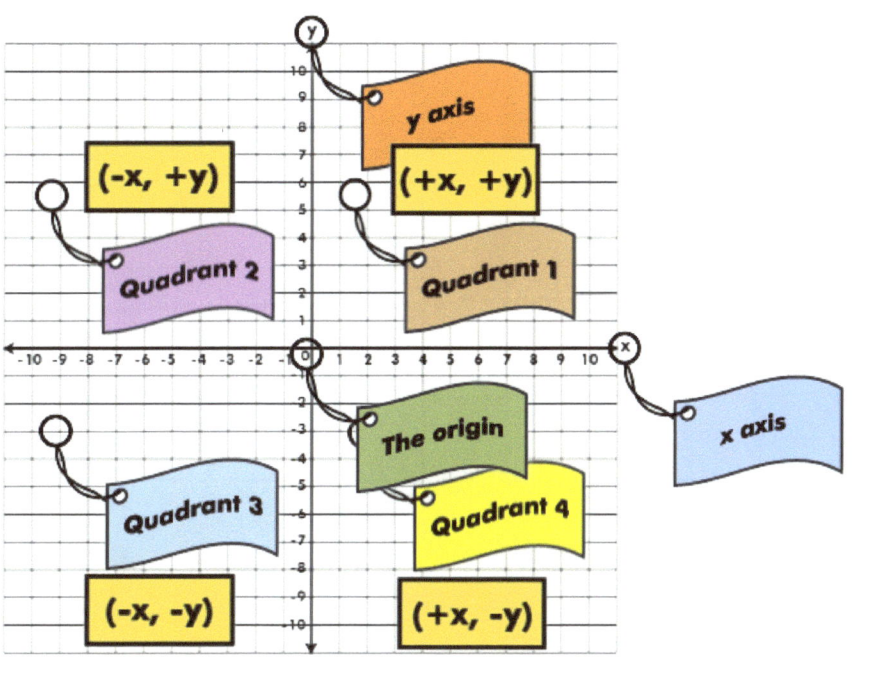

y axis

(-x, +y)   (+x, +y)

Quadrant 2   Quadrant 1

x axis

The origin

Quadrant 3   Quadrant 4

(-x, -y)   (+x, -y)

Ordered _____ **Pairs**

● (3,4)

(3,4) is the ordered pair. The x-coordinate is the first coordinate in an ordered pair and the y-coordinate is second.

Write the ordered pairs for A, B and C.

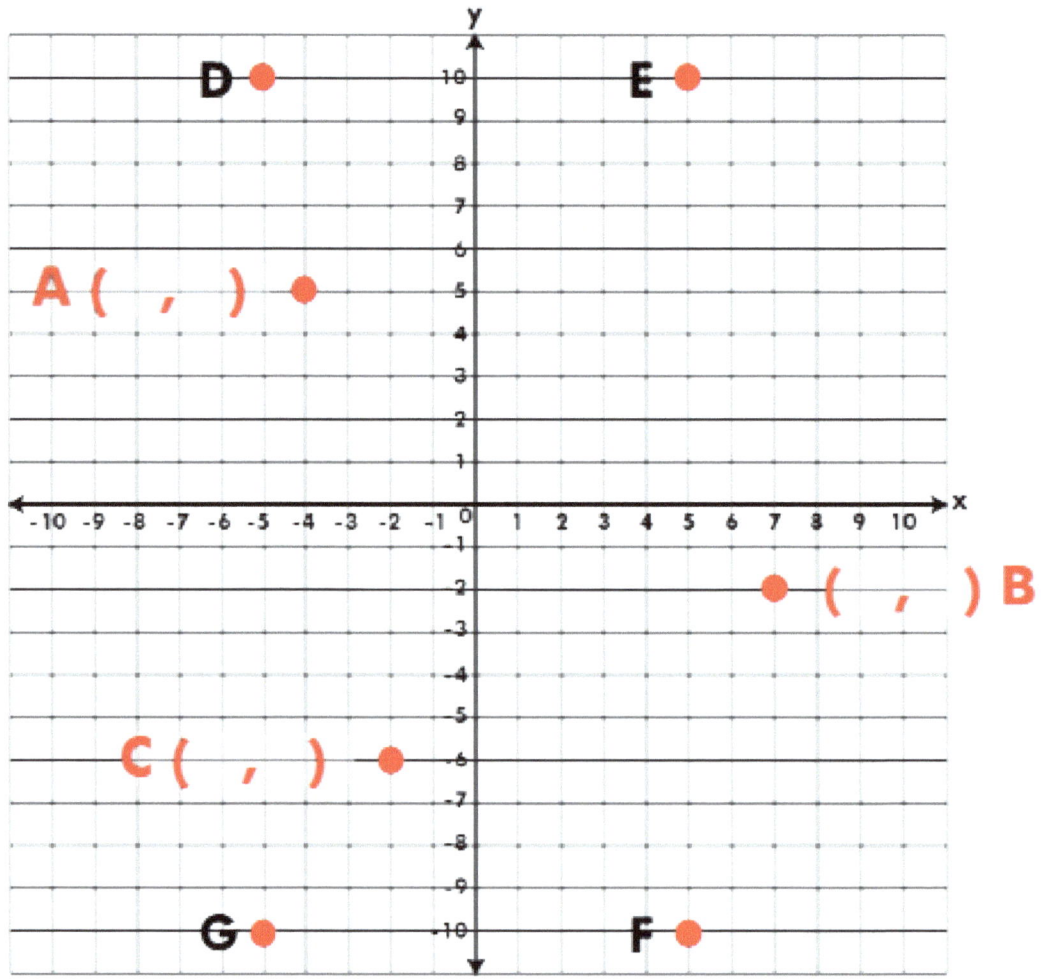

Which point has the coordinates (-5,10)?

Add and label a point to form an isosceles triangle.

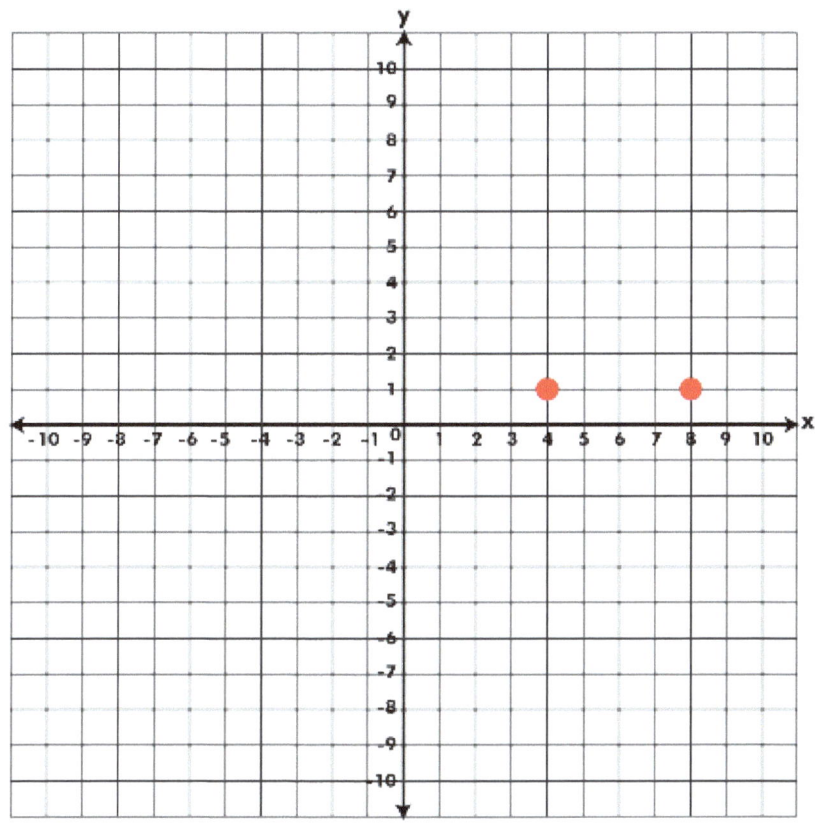

In which quadrant(s) do the points lie?

Graph the points.

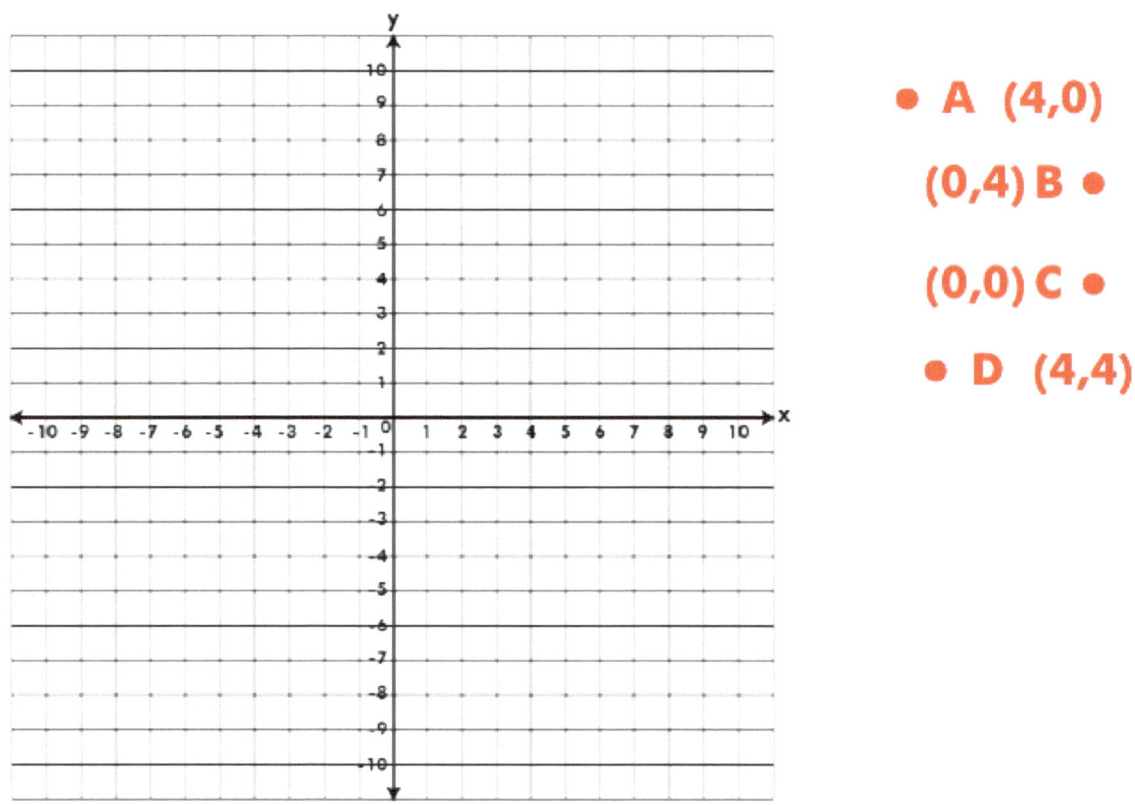

• A (4,0)

(0,4) B •

(0,0) C •

• D (4,4)

If you connect the points, what shape have you made?

Find the missing point for a hexagon.

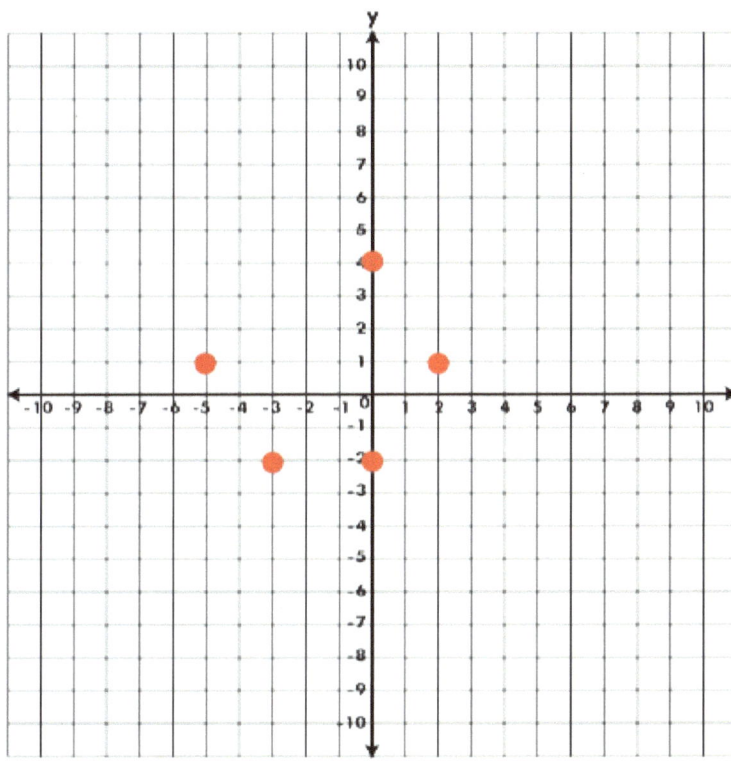

What is the point that will complete a hexagon?

(___ , ___) •

In which quadrant does the point lie?

Name_____

The Coordinate Plane Quiz

1 True or false: the ordered pair for G is (-5,5)

2 Which of the ordered pairs are located in quadrant 2?

   A (-4,1) and (6,-3)

   B (4,-1) and (-6,3)

   C (-4,-1) and (-6,-3)

   D (-4,1) and (-6,3)

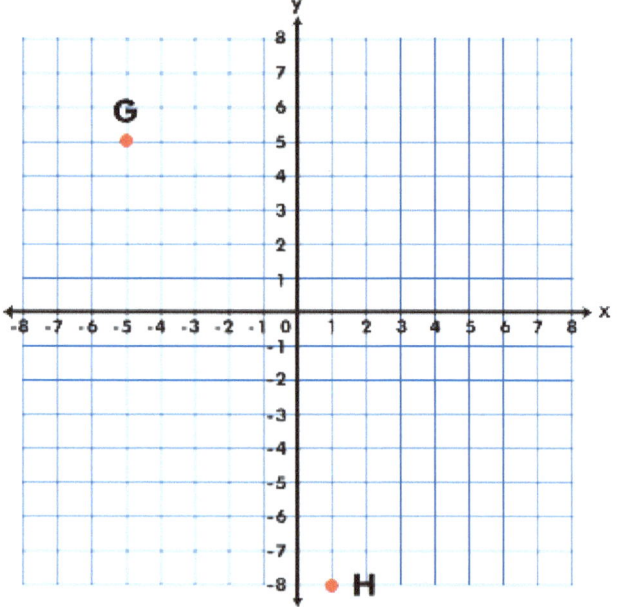

3 If FGH forms an isosceles triangle, what is the x coordinate of F?

4 If FGH forms an isosceles triangle, what is the y coordinate of F?

Transformations of the Coordinate Plane

Key Vocabulary

transformation

rotate

reflect

translate

coordinate plane

Translate triangle A per the instructions below.
Draw the triangle in the two positions as instructed. The first one is completed for you.

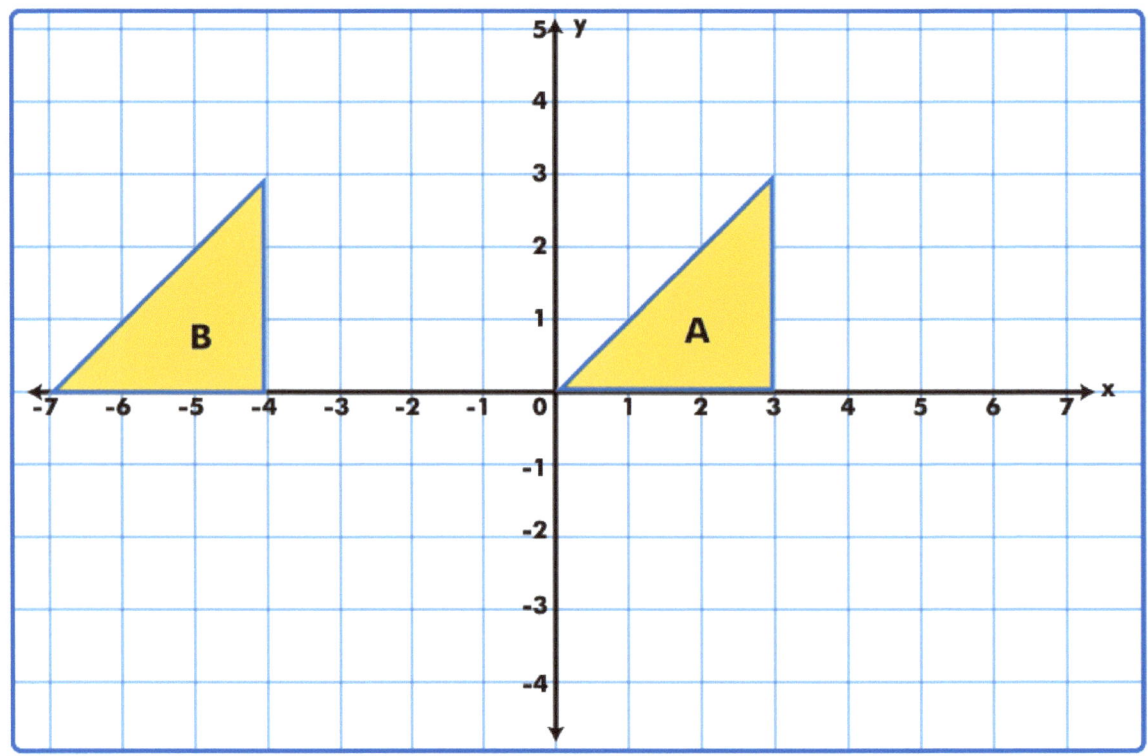

**Translate Triangle A
7 left and label as B.**

**Translate Triangle A
4 right and 1 up and
label as C.**

**Translate Triangle A
3 right and 4 down
and label as D.**

Shape A is reflected across the y-axis and labeled as B.

Study the illustration and then for the second illustration reflect shape A across the y-axis.

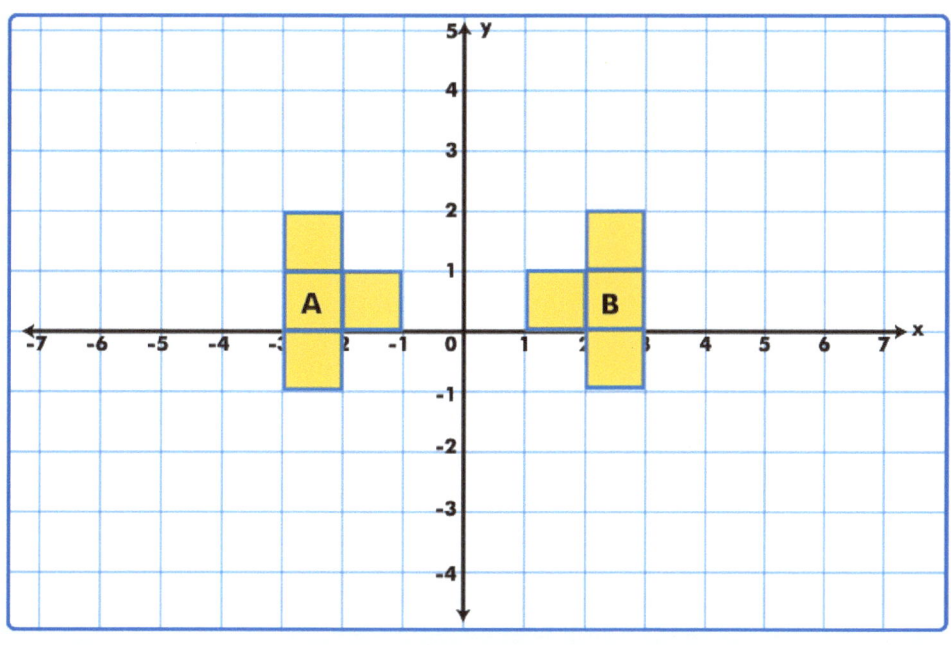

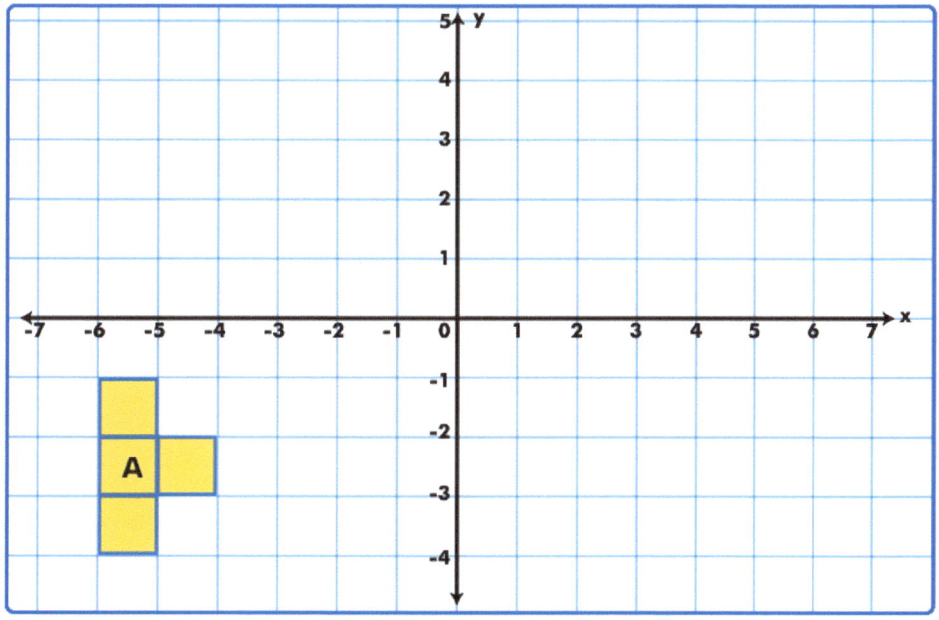

Reflect shape A across the x-axis and label it B.

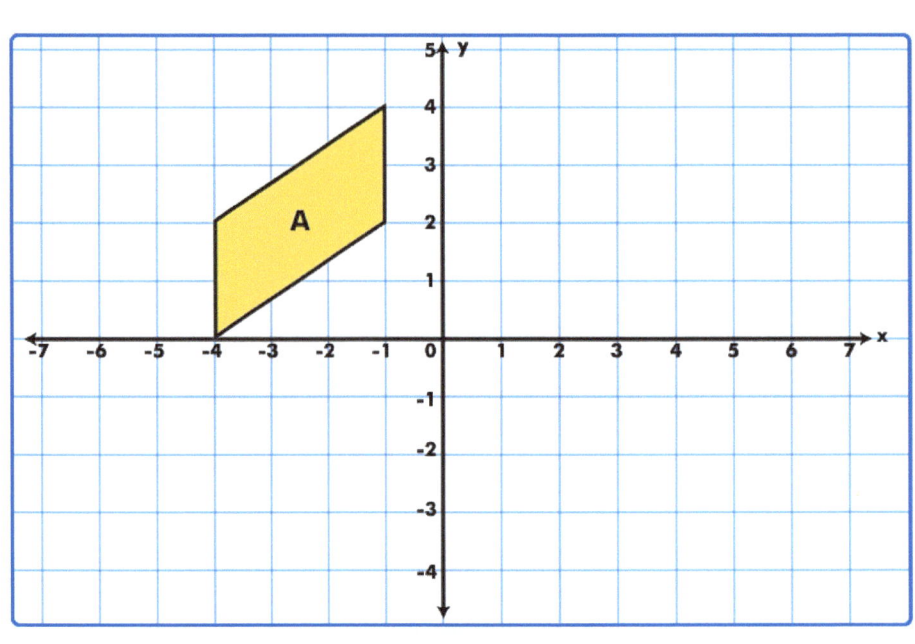

_____

Reflect and label points A, B and C across the mirror line.

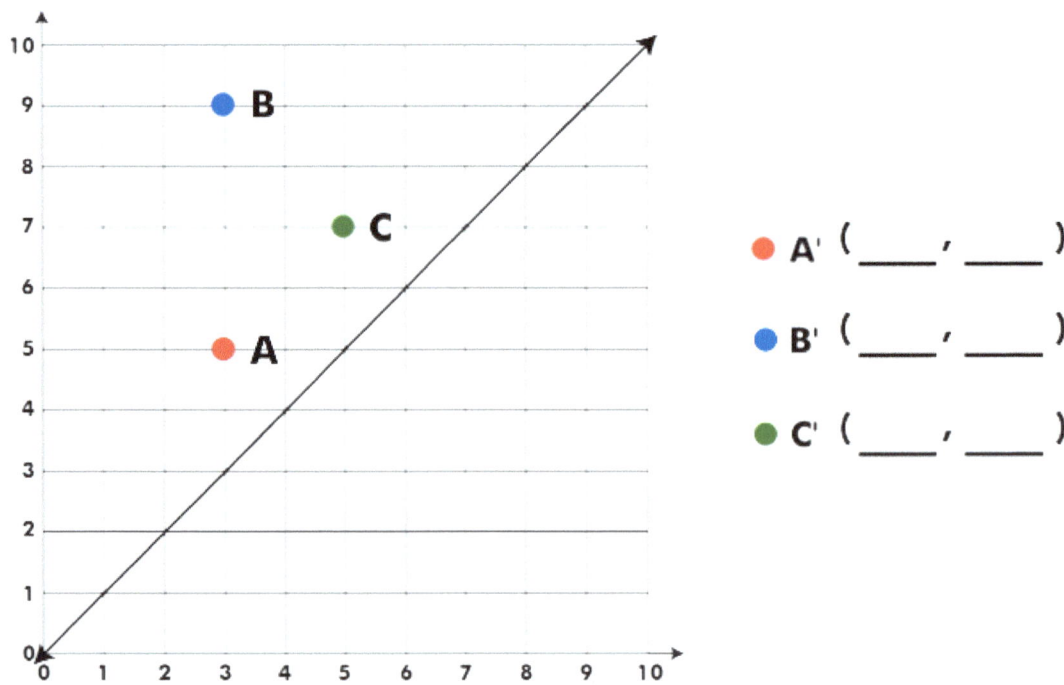

A' (_____ , _____)

B' (_____ , _____)

C' (_____ , _____)

Name_____

Transformations of the Coordinate Plane Quiz

1 True or false? If any point (x,y) is reflected across this mirror line, the coordinate pair of the reflected point will be (y,x).

2 Which image is a translation of trapezoid E?

 **A**

 **B**

 **C**

 **D**

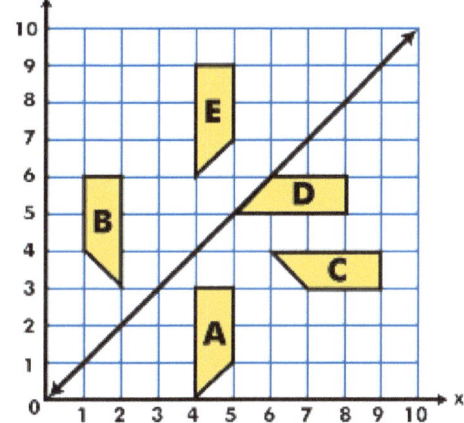

3 The translation that mapped E onto A was ___ units in the horizontal direction?

4 The translation that mapped E onto A was ___ unit in the vertical direction?

Customary Measure

Key Vocabulary

customary units

weights

capacity

Match the customary units of measure.

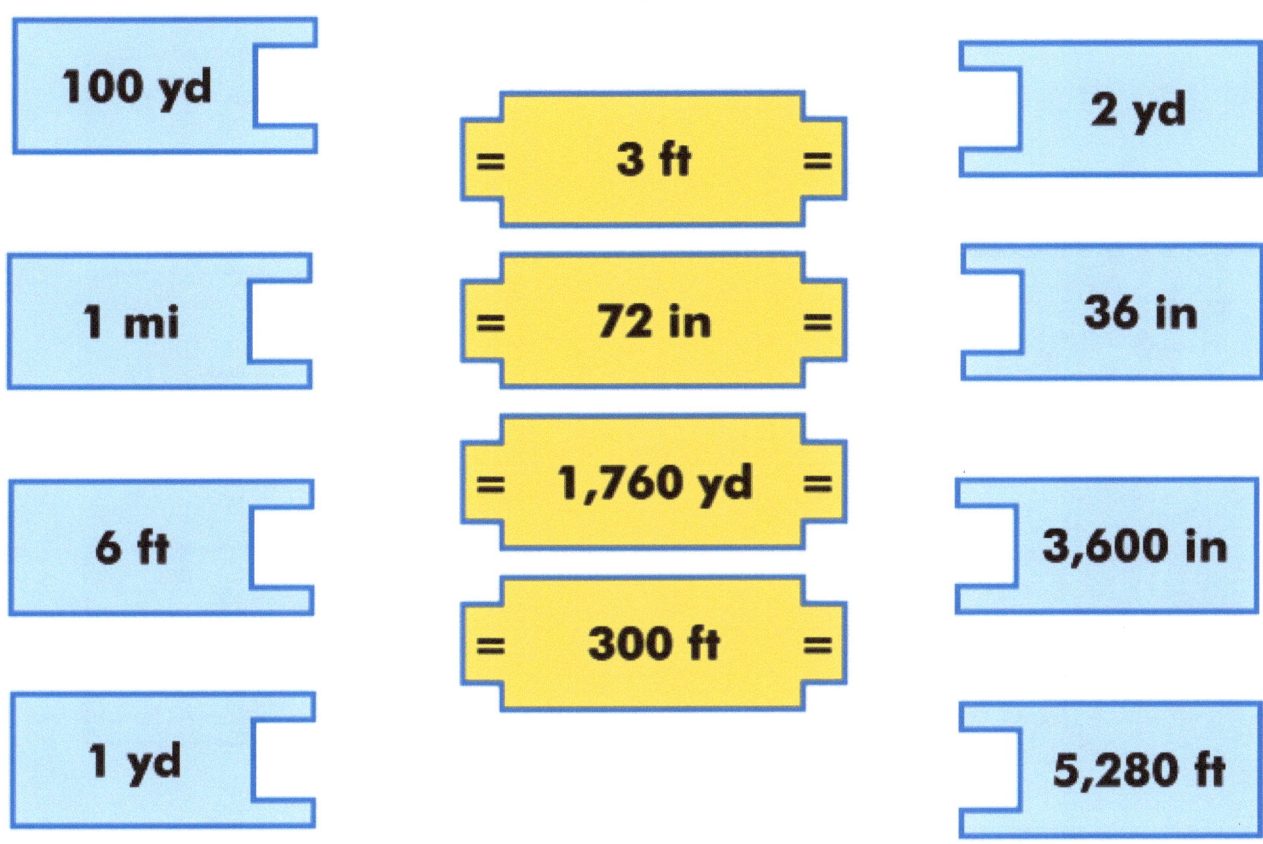

Measure the length of each item.

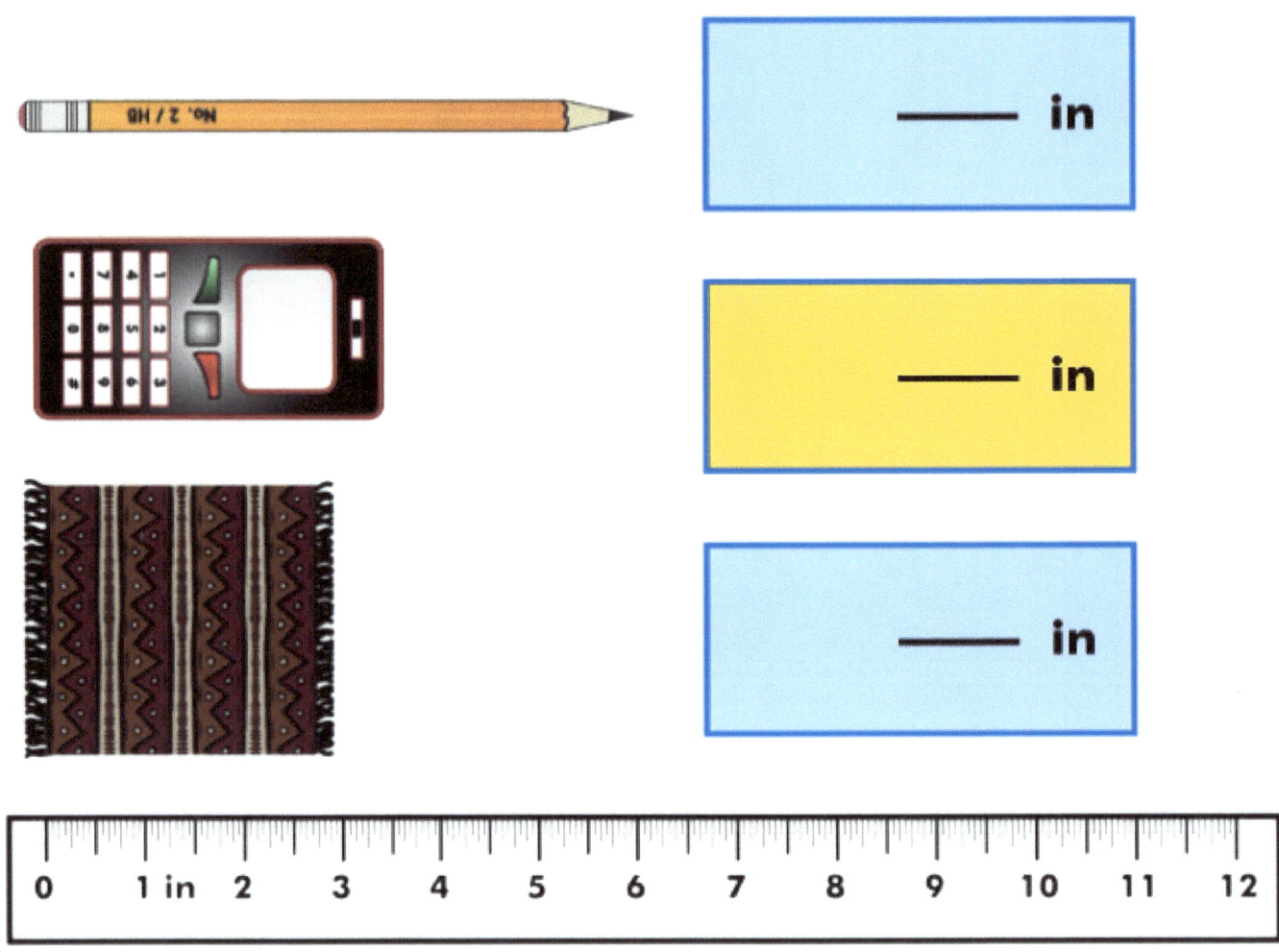

Weight

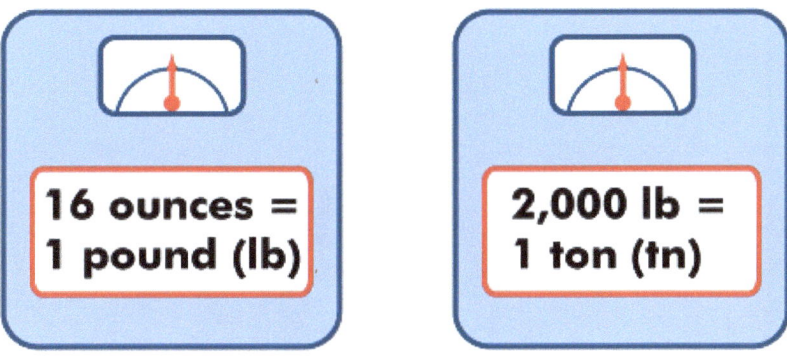

Under each scale list 3-5 things that weigh approximately the amount shown on the scale.

A few items are listed to get you started

Toast   Loaf of Bread   Small Car

How heavy is each object?
Circle the weight you estimate.

2 oz	5 oz	4 lb
6 oz	2 lb	12 oz
3.5 lb	5 lb	2 oz
8 lb	8 lb	1 lb

Capacity

1 gallon (gal) = 4 quarts (qu)
1 quart = 2 pints (pt)
1 pint = 2 cups (c)
1 cup = 8 ounces (oz)

————————————

Match the can segments.

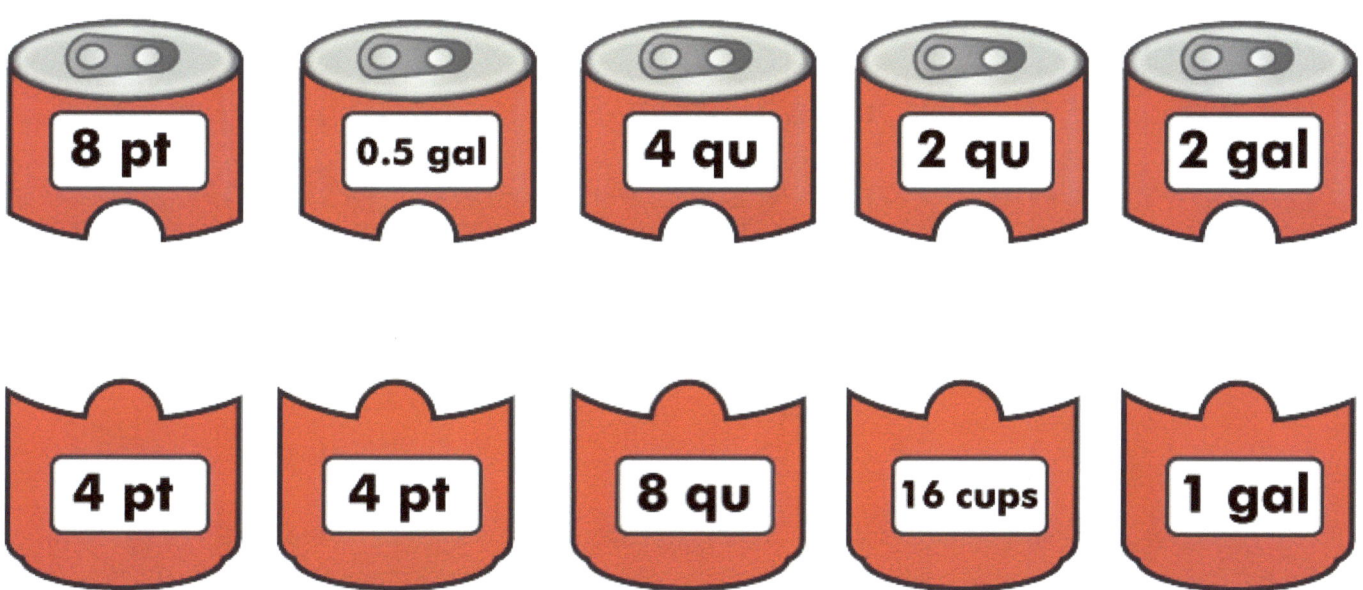

8 pt　0.5 gal　4 qu　2 qu　2 gal

4 pt　4 pt　8 qu　16 cups　1 gal

Name_____

Customary Measure Quiz

1   True or false? 1 mile = 1,762 yd

2   What is the length of this pencil?

   A 9 in

   B $9\frac{1}{4}$ in

   C $9\frac{3}{8}$ in

   D $9\frac{1}{2}$ in

3 $3\frac{1}{2}$ lb = ___ oz?

4   5 gallons = ___ pints?

Newburyport, MA 01950

1-800-596-3175

OnBoard Academics employs teachers to make lessons for teachers! We create and publish a wide range of aligned lessons in math, science and ELA for use on most EdTech devices including whiteboard, tablets, computers and pdfs for printing.

All of our lessons are aligned to the common core, the Next Generation Science Standards and all state standards.

If you like our products please visit our website for information on individual lessons, teachers licenses, building licenses, district licenses and subscriptions.

Thank you for using OnBoard Academic products.